## Preface

We are proud to present the "Science Olympiad All Chapters Combined MCQ Class 6", a comprehensive resource designed to help students excel in Science Olympiad examinations. This book provides Multiple-Choice Questions (MCQs) covering all chapters of the Class 6 science syllabus, aimed at fostering a deeper understanding of concepts and developing problem-solving skills.

Science is a subject of curiosity, innovation, and discovery. This book has been carefully designed to simplify complex concepts, engage young learners, and prepare them for the challenges of Olympiad exams. The questions are aligned with the Olympiad format to ensure students gain familiarity and confidence in tackling competitive exams.

We hope this book inspires Class 6 students to excel in Olympiad competitions, deepen their understanding of science, and develop a lifelong love for learning and discovery.

# Table of Contents

## Questions: Part 1

1. What is the primary source of energy in our diet?

a) Vitamins

b) Proteins

c) Carbohydrates

d) Fats

2. Which of the following is a natural fiber?

a) Nylon

b) Silk

c) Polyester

d) Acrylic

3. Which vitamin is important for good vision?

a) Vitamin A

b) Vitamin C

c) Vitamin D

d) Vitamin K

4. What part of the plant is the main source of sugar?

a) Root

b) Leaf

c) Stem

d) Fruit

5. Which of the following is a root vegetable?

a) Tomato

b) Carrot

c) Apple

d) Cabbage

6. Which mineral is necessary for strong bones?

a) Iron

b) Calcium

c) Potassium

d) Sodium

7. What process is used to separate cotton fibers from seeds?

a) Spinning

b) Ginning

c) Weaving

d) Knitting

8. Which of the following is not a type of cereal?

a) Wheat

b) Rice

c) Barley

d) Potato

9. Which nutrient is essential for building muscles?

a) Carbohydrates

b) Proteins

c) Fats

d) Vitamins

10. Which of these is a rich source of dietary fiber?

a) Meat

b) Rice

c) Fruits

d) Cheese

11. Which of the following is a plant-based food?

a) Milk

b) Meat

c) Bread

d) Cheese

12. What is the process called when a solid turns directly into a gas?

a) Condensation

b) Evaporation

c) Sublimation

d) Melting

13. Which component is necessary for photosynthesis in plants?

a) Chlorophyll

b) Starch

c) Oxygen

d) Nitrogen

14. Which of these is a plant fiber?

a) Wool

b) Silk

c) Jute

d) Nylon

15. What is the unit of force?

a) Joule

b) Newton

c) Watt

d) Ampere

16. What is the primary function of the roots of a plant?

a) Photosynthesis

b) Reproduction

c) Water absorption

d) Support

17. Which gas is released during photosynthesis?

a) Carbon dioxide

b) Oxygen

c) Nitrogen

d) Hydrogen

18. What is the boiling point of water at sea level?

a) 50°C

b) 75°C

c) 100°C

d) 150°C

19. What is the main function of the human heart?

a) Digestion

b) Circulation of blood

c) Breathing

d) Excretion

20. Which of the following is a conductor of electricity?

a) Plastic

b) Wood

c) Copper

d) Rubber

21. Which nutrient is known as the "body's building block"?

a) Fats

b) Proteins

c) Carbohydrates

d) Vitamins

22. What is the process of converting fibers into yarn called?

a) Weaving

b) Knitting

c) Spinning

d) Ginning

23. Which of these is not a method of separating mixtures?

a) Filtration

b) Sublimation

c) Evaporation

d) Distillation

24. What is the chemical formula for water?

a) $H_2O$

b) $CO_2$

c) $O_2$

d) $H_2SO_4$

25. Which part of the plant is responsible for photosynthesis?

a) Roots

b) Stem

c) Leaves

d) Flowers

26. Which organ in the human body is primarily responsible for detoxification?

a) Heart

b) Kidney

c) Liver

d) Lungs

27. Which of the following is an example of a carnivore?

a) Cow

b) Lion

c) Goat

d) Deer

28. What is the main purpose of carbohydrates in our diet?

a) Build muscles

b) Provide energy

c) Protect organs

d) Form hormones

29. What is the source of the fabric silk?

a) Plants

b) Silkworms

c) Sheep

d) Synthetic materials

30. Which one of the following is an insulator?

a) Copper

b) Silver

c) Aluminum

d) Plastic

31. What do plants release during respiration?

a) Oxygen

b) Nitrogen

c) Carbon dioxide

d) Hydrogen

32. What is the effect of heating on the expansion of solids?

a) Decrease in size

b) Increase in size

c) No change in size

d) Solids evaporate

33. What is the main gas found in the Earth's atmosphere?

a) Oxygen

b) Carbon dioxide

c) Nitrogen

d) Helium

34. Which food is rich in iron?

a) Rice

b) Spinach

c) Bread

d) Milk

35. What is the basic unit of life?

a) Atom

b) Cell

c) Molecule

d) Tissue

36. Which of the following is a chemical change?

a) Melting of ice

b) Rusting of iron

c) Breaking of glass

d) Dissolving sugar in water

37. What is the main purpose of proteins in the body?

a) Provide energy

b) Build and repair tissues

c) Store vitamins

d) Protect organs

38. Which part of the plant helps in reproduction?

a) Root

b) Stem

c) Leaf

d) Flower

39. What is the process of removing seeds from cotton called?

a) Spinning

b) Weaving

c) Ginning

d) Knitting

40. Which component of food helps in the formation of hemoglobin?

a) Calcium

b) Iron

c) Potassium

d) Sodium

41. Which of the following is an omnivore?

a) Lion

b) Cow

c) Bear

d) Goat

42. What is the process of making fabric from yarn called?

a) Spinning

b) Weaving

c) Ginning

d) Sewing

43. What is the main component of the Earth's crust?

a) Oxygen

b) Nitrogen

c) Silicon

d) Aluminum

44. Which gas is essential for respiration?

a) Carbon dioxide

b) Nitrogen

c) Oxygen

d) Hydrogen

45. Which of the following is a renewable source of energy?

a) Coal

b) Oil

c) Solar

d) Natural gas

46. What is the process by which plants make their food?

a) Respiration

b) Photosynthesis

c) Transpiration

d) Reproduction

47. Which organ is part of the human circulatory system?

a) Brain

b) Liver

c) Heart

d) Lungs

48. What is the hardest natural substance on Earth?

a) Gold

b) Iron

c) Diamond

d) Silver

49. Which of the following is a byproduct of photosynthesis?

a) Carbon dioxide

b) Oxygen

c) Nitrogen

d) Water

50. What type of fabric is made from sheep's wool?

a) Silk

b) Cotton

c) Wool

d) Nylon

## Questions: Part 2

51. Which of the following is a physical change?

a) Burning of wood

b) Melting of ice

c) Rusting of iron

d) Cooking of food

52. Which nutrient is essential for the formation of teeth?

a) Carbohydrate

b) Fat

c) Calcium

d) Vitamin C

53. Which nutrient is essential for the formation of bones?

a) Carbohydrate

b) Fat

c) Calcium

d) Vitamin C

54. What is the process by which water is absorbed by plant roots?

a) Transpiration

b) Photosynthesis

c) Absorption

d) Respiration

55. Which of these is an example of a herbivore?

a) Lion

b) Tiger

c) Deer

d) Eagle

56. Which of the following is a method of water purification?

a) Filtration

b) Evaporation

c) Distillation

d) All of the above

57. What is the function of chlorophyll in plants?

a) Absorb water

b) Absorb sunlight

c) Absorb nutrients

d) Absorb carbon dioxide

58. Which of these is a nonrenewable resource?

a) Wind

b) Solar

c) Coal

d) Geothermal

59. What does the liver produce that aids in digestion?

a) Insulin

b) Bile

c) Saliva

d) Gastric juice

60. Which of the following is a legume?

a) Wheat

b) Rice

c) Soybean

d) Barley

61. What is the term for an animal that only eats meat?

a) Herbivore

b) Omnivore

c) Carnivore

d) Insectivore

62. Which of these is not a conductor of electricity?

a) Copper

b) Gold

c) Wood

d) Silver

63. What is the purpose of the roots of a plant?

a) To absorb sunlight

b) To anchor the plant

c) To produce flowers

d) To reproduce

64. Which gas do animals exhale?

a) Oxygen

b) Carbon dioxide

c) Nitrogen

d) Helium

65. What is the SI unit of temperature?

a) Joule

b) Celsius

c) Kelvin

d) Fahrenheit

66. Which of these processes is used to separate salt from water?

a) Filtration

b) Distillation

c) Evaporation

d) Sedimentation

67. What is the main function of red blood cells?

a) Fighting infections

b) Clotting blood

c) Carrying oxygen

d) Digesting food

68. What does the process of pollination involve?

a) Seed formation

b) Transfer of pollen

c) Photosynthesis

d) Water absorption

69. Which of the following materials is biodegradable?

a) Plastic

b) Glass

c) Paper

d) Aluminum

70. What type of energy is stored in a battery?

a) Kinetic energy

b) Thermal energy

c) Chemical energy

d) Potential energy

71. Which organ in the human body is responsible for filtering blood?

a) Heart

b) Liver

c) Kidney

d) Lungs

72. What do you call the process of water vapor turning into liquid?

a) Evaporation

b) Condensation

c) Sublimation

d) Precipitation

73. Which of these is an example of a physical change?

a) Burning wood

b) Rusting iron

c) Freezing water

d) Baking a cake

74. Which is the largest planet in our solar system?

a) Earth

b) Mars

c) Jupiter

d) Saturn

75. What is the primary gas responsible for global warming?

a) Oxygen

b) Nitrogen

c) Carbon dioxide

d) Argon

76. What type of lens is used in a magnifying glass?

a) Concave lens

b) Convex lens

c) Cylindrical lens

d) Bifocal lens

77. Which of these organisms can perform photosynthesis?

a) Animals

b) Fungi

c) Plants

d) Bacteria

78. What is the main reason for the seasons on Earth?

a) Earth's distance from the sun

b) The tilt of Earth's axis

c) The rotation of the Earth

d) The shape of Earth's orbit

79. Which nutrient is most abundant in the human body?

a) Fat

b) Protein

c) Water

d) Carbohydrate

80. What is the smallest unit of an element?

a) Atom

b) Molecule

c) Compound

d) Cell

81. Which process is involved in the formation of clouds?

a) Condensation

b) Evaporation

c) Sublimation

d) Precipitation

82. Which of these materials is the best conductor of heat?

a) Wood

b) Plastic

c) Metal

d) Rubber

83. What type of organisms decompose dead matter?

a) Producers

b) Consumers

c) Decomposers

d) Herbivores

84. Which of these animals is known for its ability to regenerate its limbs?

a) Frog

b) Starfish

c) Lizard

d) Snail

85. What is the boiling point of water in Celsius at sea level?

a) 90°C

b) 100°C

c) 110°C

d) 120°C

86. What is the process of separating oil from water called?

a) Filtration

b) Decantation

c) Distillation

d) Evaporation

87. What is the most abundant gas in the Earth's atmosphere?

a) Oxygen

b) Carbon dioxide

c) Nitrogen

d) Hydrogen

88. Which of the following is an example of a solid?

a) Water

b) Air

c) Ice

d) Steam

89. What is the main function of white blood cells?

a) Transport oxygen

b) Fight infections

c) Store nutrients

d) Produce energy

90. Which process involves the movement of water through a plant?

a) Respiration

b) Transpiration

c) Photosynthesis

d) Pollination

91. Which of the following is a primary color?

a) Green

b) Orange

c) Blue

d) Purple

92. What is the primary function of the leaves of a plant?

a) Absorb water

b) Produce food

c) Anchor the plant

d) Protect the stem

93. Which of the following animals is a mammal?

a) Shark

b) Frog

c) Whale

d) Lizard

94. What is the chemical formula for carbon dioxide?

a) CO

b) $CO_2$

c) $O_2$

d) $H_2O$

95. What is the main function of the human digestive system?

a) Circulate blood

b) Break down food

c) Control movement

d) Produce energy

96. Which of these is a good source of Vitamin C?

a) Banana

b) Orange

c) Bread

d) Cheese

97. Which type of energy is stored in food?

a) Kinetic energy

b) Thermal energy

c) Chemical energy

d) Potential energy

98. What is the natural satellite of the Earth?

a) Sun

b) Mars

c) Moon

d) Venus

99. What is the main source of energy for life on Earth?

a) Water

b) Air

c) The Sun

d) Soil

100. What is the function of the stomata in leaves?

a) Absorb water

b) Exchange gases

c) Produce chlorophyll

d) Store nutrients

## Questions – Part 3

101. Which gas is necessary for burning?

a) Oxygen

b) Carbon dioxide

c) Nitrogen

d) Hydrogen

102. Which part of the plant is called the food factory?

a) Stem

b) Leaf

c) Root

d) Flower

103. What is the process of separating grains from chaff called?

a) Winnowing

b) Threshing

c) Sieving

d) Filtration

104. Which is the largest organ inside the human body?

a) Brain

b) Heart

c) Liver

d) Stomach

105. Which planet is known as the "Blue Planet"?

a) Earth

b) Neptune

c) Uranus

d) Venus

106. What is the basic unit of life?

a) Tissue

b) Cell

c) Organ

d) Atom

107. Which gas do plants take in during photosynthesis?

a) Oxygen

b) Carbon dioxide

c) Nitrogen

d) Hydrogen

108. Which organ in the human body helps in excretion?

a) Lungs

b) Kidneys

c) Heart

d) Stomach

109. Which is the hardest substance in the human body?

a) Bones

b) Teeth enamel

c) Nails

d) Skull

110. What is the process of water changing into water vapor called?

a) Condensation

b) Evaporation

c) Freezing

d) Melting

111. Which is the fastest land animal?

a) Tiger

b) Lion

c) Cheetah

d) Deer

112. Which vitamin is produced by the body when exposed to sunlight?

a) Vitamin A

b) Vitamin B

c) Vitamin C

d) Vitamin D

113. Which planet is known for its rings?

a) Mars

b) Jupiter

c) Saturn

d) Neptune

114. What is the instrument used to measure temperature?

a) Barometer

b) Thermometer

c) Hygrometer

d) Speedometer

115. Which part of the plant absorbs water and minerals from the soil?

a) Leaf

b) Stem

c) Root

d) Flower

116. Which is the outermost layer of the Earth?

a) Core

b) Mantle

c) Crust

d) Surface

117. Which part of the human skeleton protects the brain?

a) Ribs

b) Spine

c) Skull

d) Pelvis

118. Which organ pumps blood in the body?

a) Brain

b) Heart

c) Lungs

d) Kidneys

119. Which is the largest planet in the solar system?

a) Earth

b) Saturn

c) Jupiter

d) Neptune

120. What is the process of conversion of water into ice called?

a) Evaporation

b) Condensation

c) Freezing

d) Melting

121. Which animal lays eggs?

a) Cow

b) Hen

c) Dog

d) Cat

122. What is the main source of energy for the Earth?

a) Moon

b) Sun

c) Wind

d) Coal

123. Which type of rock is formed by cooling of molten lava?

a) Igneous rock

b) Sedimentary rock

c) Metamorphic rock

d) Fossil rock

124. Which part of the digestive system absorbs most nutrients?

a) Stomach

b) Small intestine

c) Large intestine

d) Liver

125. Which blood cells help in clotting of blood?

a) Red blood cells

b) White blood cells

c) Platelets

d) Plasma

126. Which force makes objects fall to the ground?

a) Friction

b) Gravity

c) Magnetism

d) Air pressure

127. Which is the main gas in the Earth's atmosphere?

a) Oxygen

b) Nitrogen

c) Carbon dioxide

d) Helium

128. Which type of simple machine is a knife?

a) Lever

b) Inclined plane

c) Wedge

d) Pulley

129. What is the process of conversion of water vapor into liquid water called?

a) Evaporation

b) Condensation

c) Freezing

d) Melting

130. Which organ controls all the activities of the body?

a) Heart

b) Brain

c) Lungs

d) Kidneys

## Answers

1. c) Carbohydrates

2. b) Silk

3. a) Vitamin A

4. d) Fruit

5. b) Carrot

6. b) Calcium

7. b) Ginning

8. d) Potato

9. b) Proteins

10. c) Fruits

11. c) Bread

12. c) Sublimation

13. a) Chlorophyll

14. c) Jute

15. b) Newton

16. c) Water absorption

17. b) Oxygen

18. c) 100°C

19. b) Circulation of blood

20. c) Copper

21. b) Proteins

22. c) Spinning

23. b) Sublimation

24. a) H2O

25. c) Leaves

26. c) Liver

27. b) Lion

28. b) Provide energy

29. b) Silkworms

30. d) Plastic

31. c) Carbon dioxide

32. b) Increase in size

33. c) Nitrogen

34. b) Spinach

35. b) Cell

36. b) Rusting of iron

37. b) Build and repair tissues

38. d) Flower

39. c) Ginning

40. b) Iron

41. c) Bear

42. b) Weaving

43. c) Silicon

44. c) Oxygen

45. c) Solar

46. b) Photosynthesis

47. c) Heart

48. c) Diamond

49. b) Oxygen

50. c) Wool

51. b) Melting of ice

52. c) Calcium

53. c) Calcium

54. c) Absorption

55. c) Deer

56. d) All of the above

57. b) Absorb sunlight

58. c) Coal

59. b) Bile

60. c) Soybean

61. c) Carnivore

62. c) Wood

63. b) To anchor the plant

64. b) Carbon dioxide

65. c) Kelvin

66. c) Evaporation

67. c) Carrying oxygen

68. b) Transfer of pollen

69. c) Paper

70. c) Chemical energy

71. c) Kidney

72. b) Condensation

73. c) Freezing water

74. c) Jupiter

75. c) Carbon dioxide

76. b) Convex lens

77. c) Plants

78. b) The tilt of Earth's axis

79. c) Water

80. a) Atom

81. a) Condensation

82. c) Metal

83. c) Decomposers

84. b) Starfish

85. b) 100°C

86. b) Decantation

87. c) Nitrogen

88. c) Ice

89. b) Fight infections

90. b) Transpiration

91. c) Blue

92. b) Produce food

93. c) Whale

94. b) CO2

95. b) Break down food

96. b) Orange

97. c) Chemical energy

98. c) Moon

99. c) The Sun

100. b) Exchange gases

101. a) Oxygen
102. b) Leaf
103. a) Winnowing
104. c) Liver
105. a) Earth
106. b) Cell
107. b) Carbon dioxide
108. b) Kidneys
109. b) Teeth enamel
110. b) Evaporation
111. c) Cheetah
112. d) Vitamin D
113. c) Saturn
114. b) Thermometer
115. c) Root
116. c) Crust
117. c) Skull
118. b) Heart
119. c) Jupiter
120. c) Freezing
121. b) Hen
122. b) Sun
123. a) Igneous rock
124. b) Small intestine
125. c) Platelets
126. b) Gravity
127. b) Nitrogen
128. c) Wedge
129. b) Condensation
130. b) Brain

Printed at Repro India Ltd.